KEEPING
A COUPLE OF PIGS
IN YOUR GARDEN

A Kitchen Garden Book

Text © Karen Nethercott
Illustrations © Gabrielle Stoddart

In the same series:
Keeping a Few Hens in your Garden
Food from the Kitchen Garden (with P Hands)
Keeping a few Ducks in your Garden
A Peacock on the Lawn (with S Carpenter)
Goose on the Green
Start your own Garden Farm (with K Nethercott)
Also:
The Big Book of Garden Hens
A Henkeeper's Journal
A Christmas Journal
All my Eggs in one Basket

Published by the Kitchen Garden
Troston Suffolk IP31 1EX
Tel 01359 268 322
Email:francine@kitchen-garden-hens.co.uk
www.kitchen-garden-hens.co.uk

ISBN 978-0-9561952-1-0
Printed in England on re-cycled paper

"A cat will look down to a man. A dog will look up to a man. But a pig will look you straight in the eye and see his equal"

Winston Churchill

Introduction

Without doubt, of all the adventures we've had over the past few years as novice smallholders, the one I've enjoyed the most has been keeping pigs. From the moment we unloaded our first squealing weaners, to the pleasurable hours spent six months later, making salamis from those same animals, I've loved every minute. Not only did I get to know and fall in love with my pigs – the most characterful of all farm animals, but I learned a multitude of new culinary skills; making ham, bacon and sausages, and even dabbled in a bit of *charkuterie*, ending up with a freezer full of the most delicious meat I've ever tasted.

If you are considering having a go yourself, then I'd recommend you become a 'fair weather' pig keeper initially. Buy a couple of weaners in the spring, keep them over the summer, and take them to slaughter in the autumn. You and your pigs can enjoy those lazy, hazy days; your ground won't turn into a swamp, and your larder will be stuffed with goodies in time for Christmas.

Don't worry if you haven't got perfect accommodation, pigs are very forgiving. Our first pig area was a long thin strip down the side of a field. The following year, we fenced off a corner of the garden that needed clearing of brambles. Pigs don't need a huge amount of space, but as a guide I'd aim for at least 400 sq meters for a pair of weaners. Never keep a single pig, it just isn't fair – they love the company of others, so they can get up to mischief together.

My main aim, in this book, is to share my experiences with you, and hopefully inspire you with enough confidence to have a go yourself. Yes, you will miss your piggies when they go, but if you eat pork, surely it must be better to eat animals you know have lived a really good life?

Where to begin

If you intend to keep pigs for food, your first task is to locate the nearest abattoir that will accept pigs as private kills. Most areas are well-served, and you'll have a choice, but some counties are not so well covered, and this may be a problem. My journey time to the abattoir here in Norfolk is just over half an hour. I'd probably still keep pigs if we had to travel for an hour, but no further.

If you live in a built-up area, it would be worth sounding out the local Environmental Health Office to make sure they don't foresee any problems, and check if your property deeds prohibit the keeping of pigs. It's a good idea to let your neighbours know too. Most people love seeing outdoor pigs, so you're more likely to be deluged with requests for visits, rather than objections.

Next, contact DEFRA to obtain a holding number and a herd number (their website address is at the back of this book). Don't be put off by the bureaucracy, persevere – you only need to do it once. Then call your local Trading Standards, and ask them to send you some blank movement licences and a movements book. Also check what records they want you to keep, because you may get an inspection in future.

While you're waiting for the paperwork to turn up, you can have fun looking for your first piglets. I have always chosen to keep rare breed pigs. They're slow-growing, well-suited to outdoor life, and definitely produce the best-tasting pork. The Rare Breed Survival Trust's website has a complete list of available breeds, along with their society contacts. The Breed Clubs will be able to help you find local herds who may have weaners for sale. To help narrow down the choice, I've included my Top Six Breeds on page 16.

I found my first weaners through my local smallholding group, and my breeding animals through the British Saddleback Club. Try to go to a reputable source, avoid the local free ads, and if you want pedigree pigs, remember to check their registration forms. The breed clubs will tell you how much to expect to pay for a weaner, but a lot will depend on local supply and demand. A rough guide is about £5 per week, so a pedigree 10 week old weaner should cost approximately £50.

Always visit the litter before you buy, and be prepared to be seduced by the piglets. I must have spent an hour deciding which ones to choose. They were Gloucester Old Spots, one of my favourite breeds, and in the end I chose the two spottiest girls. Make sure your choices are the same sex. The last thing you want is to take a pregnant pig to slaughter.

We must have made every mistake in the book with our first charges. Thank goodness no-one was watching when we brought them home. We'd set up the paddock in advance, bought a pig ark and filled it with straw, encircling it with 4 strands of electric fencing. Our plan was to back the trailer up to the paddock, and then pick up each pig and carry it over to the ark.

Jeff picked up the first piglet – and quickly dropped it. How could such a small animal make a noise like that? He tried again, and heard the same horrendous screaming. I could almost see the sound waves rippling the air around him. Not to be outdone, he masterfully picked up one of the girls, and dumped her into the paddock. She immediately took off, and ran right through the electric fencing. Half an hour later we managed to catch her and put her into the pig ark. While I stood guard at the entrance, Jeff went and got her sister. Carrying the screaming piglet at arm's length, facing away from him, I could almost see his ears bleed (I had my fingers in mine). At last, both were in their house, snuggled down together under the straw, asleep. Over the next couple of days, after more escapes and alterations to fencing arrangements, we finally got it right. We learned valuable lessons, and from then on always replicated the type of fencing the piglets were used to. Oh and, if we can avoid it, we never pick up a piglet!

Young pigs are utterly captivating, and you and your family will probably become very attached to these charming creatures. As a result, you may find it daunting when faced with 'sending them off'. I was in tears the first time, but it has subsequently got easier. Just bear in mind why you are actually keeping them. Children are usually far more pragmatic than adults. A friend's kids actually named their piglets: 'Bacon' and 'Sausages' – much to her horror. Remember: your pigs arrive as adorable toddlers, but leave as belligerent teenagers! And they can bite, not usually out of malice but just in play, and there's no reason why children can't enjoy helping with your pigs under supervision.

Finally, remember your new charges must be fed twice a day, have ready access to fresh water and adequate shelter from the weather. While these tasks only take a few minutes a day, you'll be tempted to spend a lot more time with them. Hosing down a couple of pigs on a sunny afternoon is one of life's greatest pleasures.

Where To Keep Your Pigs

The perfect pig paddock would be on slightly sloping, sandy soil, conveniently dotted with a few trees for shade and scratching. If these trees then provide an edible harvest that gently drops its crop over a couple of months – pig heaven. Sadly, all I have to offer is loam over clay; flat and very wet in the winter, baking hard in the summer, and almost treeless. Not ideal, but with a bit of thought, the creative pig keeper can add features to enhance even the dullest environment.

A boring patch of pasture can be made more pig-friendly by adding a large chunk of tree or a tractor tyre, and a tarpaulin can be rigged up to provide a little shade in the summer. Whatever you supply must be sturdy – you'll be amazed at a pig's strength. If you keep adult pigs, then any object within their reach will be severely tested. Our boar Freddie has been known to move his entire house across the field in an ecstasy of scratching. Fine, up until the point where the metal ark meets the electric fence, and then his true nature is revealed, as he runs off shrieking, and we have to go and rescue him!

Give your pigs as much space as you can spare – at least 400 sq meters for two weaners, and never keep a single pig. If you intend to breed, allow at least 400 sq metres per adult pig, and make sure you have the same again, lying empty, so that you can rotate paddocks.

A wallow in summer is really important. Pigs don't sweat and easily over-heat, so they need to cover themselves in mud, which not only cools them down, but acts as an effective sunscreen, so they don't burn. Usually, you don't have to make anything for them, just let them make their own wallow by tipping water from their drinking trough, or run a hosepipe to soak the ground and keep it topped up every few days, as they hollow out a bowl. You'll always notice depressions caused by wallowing on any ground left by pigs. Fill them in or leave them for the next tenants.

Build a basic sturdy shelter where your stock can sleep and rest. I like traditional corrugated-iron mobile pig arks. Some of ours are homemade and others have been bought second-hand. These are generally sturdier with insulated roofs, keeping their inhabitants cool in the summer and warm when it's cold. We also made a wooden one based on the dimensions of an 8′ x 4′ (2.4m x 1.2m) sheet of heavy-duty ply that worked well, but was too heavy to move around without dismantling.

When positioning the house, keep the entrance away from the prevailing wind, and bed the inside down with a good layer of straw, especially in winter. Pigs won't soil their house, but straw gets dirty especially if it's muddy outside, so keep it topped up. Your pigs will enjoy a top-up of new straw, wandering in and out of the house with mouthfuls, as they re-arrange their beds. When the new bedding arrives, our two sows chuck Freddie out of the house, so they can do their housekeeping in peace. He stands waiting patiently outside until they're ready, and he's let back in. Pig manure and bedding make excellent plant food and soil conditioner, stacked in layers on your compost heaps.

Feeders come in all shapes and sizes, and we've tried most of them over the years, but nowadays we don't use troughs for feeding. It's easier to feed straight on the ground, spreading the feed further so the greediest pigs don't grab all the food.

We've tried traditional Mexican hats, plumbed-in water troughs, and cut-down water butts as drinkers. All fine and dandy till your lovely pig turns them over with a flick of its nose. If it isn't empty, it'll be full of mud, so you will need to tip the mud out regularly and refill. If you end up keeping several pigs in a number of paddocks, then it's worth investing in a nipple drinking system.

Available from agricultural merchants, it consists of a barrel with metal nipples screwed into the side. The barrel is filled with water, either manually or by attaching a hosepipe to a ballcock system in the top, and the pigs drink from the nipples – easier for us, and ideal for the pigs, who always have clean fresh water on tap. The barrel stands on a sheet of ply, to deter them from making a wallow and undermining the barrel.

Finally, to fence your pigs, use either fixed fencing or electric wire, for which you'll need a sturdy post in each corner, and a couple of strands of wire tensioned between them. Plastic posts pushed in every few feet will keep it level. It's worth investing in a system where you wind the wire onto a reel, then use the wheel for tension. Make sure your charger is powerful enough to cover the distance, and more importantly, will still work if grass and clods of mud are chucked up onto it. The pigs will regularly try and bury the wire, so walk the perimeter on wet mornings and kick off any mud you find. Buy all your kit from your local agricultural merchant, who will advise you.

Which Breed

Most pigs you see in commercial units are prick-eared whites. Fifty years of selective breeding has created an animal that produces lots of uniform piglets, early to wean and quick to fatten. Great for *cheap* meat production, not so good for *tasty* meat.

I used to feel inadequate in the kitchen. I couldn't cook pork. Like most people, I bought my meat from the supermarket, and time and again the resulting meal was disappointing; dry, tasteless meat with rubbery crackling. In the end we stopped eating it. Then I was introduced to rare breed pork – a revelation – cooked in the same way, but the result was quite different; moist tender meat with crunchy crackling.

At the beginning of the 20th century, Britain had various distinct regional pig breeds, including the Lincolnshire Curly Coated and the Dorset Gold Tip. In the past, breeders had selectively bred from local pigs creating regional differences to suit local conditions. Some focussed exclusively on the rangy lop-eared indigenous animals, while others added Asian blood from chubby, prick-eared pigs introduced in the late 18th century.

These breeds were dominant until the 1950's, when the government's Howitt Report decided that diversity was a handicap to the development of the British Pig Industry, and recommended focusing on fewer breeds, ideally just one. Intensive factory farming flourished, and the traditional breeds were no longer necessary. Without a market for their meat, there was little point in breeding these animals, and numbers fell dramatically to the point where several became extinct. The Lincolnshire Curly Coated and the Dorset Gold Tip were lost within 15 years, and by the early 1970's most breeds were at dangerously low levels. The turning point came in 1973, when the Rare Breed Survival Trust was set up, arresting the decline and preserving this genetic heritage. Since its establishment, not a single breed of British farm animal has become extinct. The work of both the Trust and hard-working breed clubs has brought these lovely animals back from the brink of extinction, and numbers have been increasing.

By choosing rare breed pigs, you become part of this success story. Try one breed now, and another next year, and you'll notice the difference in looks, temperament and above all, in the taste and texture of the meat. You'll soon have a favourite and then, who knows, you may even want to start your own mini breeding herd. Encouragement to eat rare breeds may seem a contradiction in terms, but it's only by eating them that numbers and consequently, the genetic pool, will increase.

I'd like to warn you about a worrying trend of breeding miniature versions of pigs, marketed as pets. Sold for hundreds of pounds as teeny piglets, I hope their new owners aren't in for a shock. Adult pigs of any size make poor house guests.

These are my Top Six Breeds:
British Saddleback

Resulting from the amalgamation in 1967 of two similarly coloured breeds, the Essex and the Wessex, the British Saddleback is a problem-free hardy pig and excellent mum. We chose these friendly pigs to start our mini breeding herd - they're good for both fresh pork and bacon. Being lop-eared, they are easy to manage and even train. My sows come running when I call their names.

Gloucester Old Spot

Known as the Orchard pig, their spots are said to have been caused by falling apples! Hardy, well-suited to outdoor life, the friendliest and most docile pigs I've ever kept, but their predominantly white skin is prone to sunburn, so remember to keep them supplied with a muddy wallow in the summer. An ideal starter pig producing excellent fresh pork and bacon.

Tamworth

Originating in the Midlands, the only British red pig and the breed least 'improved' by the Asian pigs imported in the 18th century. Retaining the long snout of the earliest British types, Tamworths are thought to be the truest indigenous breed. This prick-eared pig is a mischievous handful. Good bacon pig due to its length.

Large Black

Britain's only black pig, this lop-eared breed is friendly and very docile. They make good mothers and have large litter sizes. Their numbers declined dramatically in the 1960's, due to dislike of its dark colouring, but nowadays the meat is valued for its succulence. No worries about sunburn here!

Middle White

I know lots of people love them, but I think they're the ugliest pigs in the farmyard. Though small, this white pig with pug nose and prick ears matures quickly, and has a reputation for excellent quality pork. Nowadays, the breed has a number of high profile fans among celebrity chefs and top class restaurants, all singing the praises of its succulent sweet-tasting pork.

Kune Kune

Kune Kunes (pronounced cooney cooney) are not a native breed, but these delightful chubby animals are a great option for people who want to keep pigs as pets, as opposed to the table. Small (24″-30″/60cm-75cm high), they come in a range of colours. Friendly, placid creatures that love human company, being small, they do less damage to your garden, and usually spend their time grazing, rather than rooting.

What to Feed your Pigs

Pigs will thrive on any surplus produce from the vegetable garden or orchard. We even give ours grass clippings, which they eat, play with and then finally fall asleep on. They'll consume almost anything you put in front of them, so make sure it's safe to eat. Just remember the golden rule – absolutely NO MEAT.

While vegetable bonuses will be greatly enjoyed by your pigs, the easiest way for you to ensure they get a good balanced diet is to buy a quality pelleted pig food (pignuts). Sold in 20kg bags from agricultural feed merchants, they come in a variety of shapes and sizes. Try to avoid those with fishmeal, the flavour may come through to the pork. When we first started with pigs, we bought separate weaner, grower and finisher rations, each designed to give pigs what they need at every stage of life.

Later, when we started our breeding herd, we found the piglets eating mum's sow rations after a couple of weeks, so we kept them on the same food after they were weaned. As a result, we didn't have to buy in small amounts of different rations, and could get better discount by buying in bulk. The pigs don't suffer, or have to cope with stressful changes of food and, because it has a lower protein content, it's actually better for them, as they don't put on quite as much fat in the last few weeks.

A 20kg bag of pig food will feed two weaners for two weeks; at 5-6 months old, it will last half that time. The more 'extras' you feed them, the more you can reduce the pig food. Assume vegetables are worth half the pignuts, so offer a kilo of veg and you can reduce their pellets by half a kilo. Feed twice a day and scatter the pellets on the ground, so they have fun searching for them.

Meat

No-one enjoys taking their pigs on their last journey, so focus on making it as stress-free as possible for your animals. You've given them a great life, now make sure they have a good death. The first time will be the most distressing. It will get easier from then on.

I was too upset to accompany my first pigs to the abattoir, so my husband took them instead. It's important to send them with someone they're used to – a friendly face at the other end. Load them effortlessly by putting your trailer ramp into their paddock a couple of days before, and feed them inside. Having got used to it, they'll probably be in there waiting for you on their last day. All you need to do is shut the gate behind them. Give them some straw, and throw in a few pignuts to keep them busy.

Mark your stock with your herd number, using metal ear tags or slapmark them on each shoulder on the day. We prefer to slapmark, because pigs can lose their ear tags in the rough and tumble of growing up, and it's easier to slap them in the trailer than try to attach an ear tag. They may grumble, but it's over in a second. Identification equipment is available from Ketchums in Surrey (details on page 32). Take your completed movement licence with you to the abattoir, and fill out the abattoir's Food Chain Information (FCI) form.

It's no longer practicable to get your pigs slaughtered at home, although you can still butcher the carkass yourself, if all the meat is for your own consumption. The more usual route is to arrange for either the abattoir or a local butcher to cut the meat up to your specifications. Make sure you find a butcher you trust to return all your meat to you. Incidentally, don't worry that dark pigs' meat has unusual coloured skin. All pigs have white skin by the end of the slaughter process, the only remnant of their original colour may be a few bristles.

Before deciding on your butcher, check his range. Fresh pork cuts are fine, and all butchers make their own sausages, but not all cure ham and bacon themselves. This could be the spur you need to learn to do it yourself. I enrolled on a sausage-making course and enjoyed learning to make my own sausages and salamis, and of course, eating them!

Piglets

Having mastered the art of raising weaners, you may decide to take one step further into the world of breeding. More space and a lot more time are pre-requisites, but if you have the confidence, it's great fun. Two facts may put you off: adult pigs can live for 10-15 years, so consider how you'll manage their reducing fertility; and consider also the 20 or so piglets each sow produces every year. Will you be able to eat or sell every litter? If we didn't have the shop to sell our meat, I would never have started a breeding herd.

Starting small is the best way, just two sows and maybe a boar. I say maybe, because you have four options when it comes to fathering the litters. One to keep your own boar, a second to send your sows out to a boar, three - borrow a boar, or finally AI (Artificial Insemination). Each option has pros and cons, so decide what suits you. If you keep a boar, you need to be confident you can handle him when he grows big with tusks. In the paddock, I am completely happy to turn my back on my girls, but I always keep an eye on Freddie. Though not malicious, he can do a lot of damage just by turning his head. A friend of mine squashed her finger between the doorway and her boar's shoulder – he just wanted a scratch and her hand was in the way. My husband's leg was injured when Fred turned, and inadvertently caught him with his tusk.

Sending your girls away or borrowing a boar may be simpler, if you can find one you're happy with. You'll need to consider the bio-security risk, and the stress involved in introducing new animals both to each other, and to a new environment. Going down the AI route is trickier than letting nature take its course, and only succeeds when you get the timing absolutely right. You must know exactly when your sow comes into season, order your semen at the right time, and then inseminate her at the right point in her cycle. Not for the faint-hearted, but obviates the need for a boar.

Acquire your sows as weaners. I suggest you buy four sisters, then send two away to slaughter and keep your favourites as future breeding stock. Learn to recognize temperament and good breed qualities, as these traits in the sows will be passed on to their piglets. If possible, choose girls that have at least 12 well-spaced nipples – fewer and they'll have problems feeding larger litters.

Sows are pregnant for 3 months, 3 weeks and 3 days, so check your diary before you start to make sure you won't be away when they are due. Keep their routine stable while they're pregnant. I even leave them in with Freddie until the last few days. Then they're moved over to a farrowing house with a lot more room and separate *en suite* facilities. I try to let both mums farrow together, so if necessary, I can mix the litters and keep them in single sex paddocks to get similar numbers in each.

A good reference book on farrowing is essential. I recommend Carol Harris's 'A Guide to Traditional Pig Keeping' with detailed info about the process, and what to look out for at each stage. Sometimes you just need a bit of reassurance, so cultivate a few experienced pig-friends, in case you need them. During my first couple of farrowings, I spent the procedure on the phone to an experienced breeder, describing what was happening and asking if it was normal.

The classic signs of an imminent birth, such as nest building, restlessness and milk production all happen during the last few hours. Then the sow will settle and piglets will arrive one by one. I usually bring in a straw bale and sit with her. As each piglet appears, I clear its mouth and nose, give it a rub down with a towel and settle it down by the nipples to find its own teat. Piglets lay claim to a particular nipple and return time and again – nature's way of ensuring vital energy isn't lost in an hourly scramble for positions. Farrowing can take a couple of hours, or half a day with long gaps in between. You won't know it's all over until the final afterbirth is expelled. Encourage mum up for a drink and something to eat, and leave them all to it.

Piglets are adorable. Spending time with them when they're young will make them easier to handle as they grow. It's no hardship to watch them play tag and scamper around like puppies, just be careful not to upset mum, who may be protective with her first litter. There is no need at all to castrate or tail dock, both procedures smallholders consider to be unnecessary and cruel.

Your piglets will grow at an alarming rate, doubling their birth weight in a week and munching on mum's food. She will gradually wean then off her milk by spending increasing amounts of time lying on her tummy and denying them access to her teats. I wean them fully by taking them away from their mother when they are between 8-10 weeks, although you can remove them earlier if the sow is losing condition with a large and demanding litter.

Problems

Worms can be debilitating if not dealt with promptly. Piglets are normally wormed at weaning, so check this has been done when you buy, or do it yourself when you get them home. This should be all the medication they will need during their stay. Breeding animals need regular worming unless you can provide clean grazing. The easiest way to administer wormer is in a sandwich (they'll never guess). Remember to record any drugs you administer in a notebook.

The combination of a pig's ballerina points and uneven ground can cause the occasional limp. It's unlikely to be a problem, and always seems to sort itself out in the end. Adult pigs' feet and boars' tusks may need to be trimmed if they become too long. Pigs that are used to being handled may let you trim their feet. My friend has pet Kune Kunes and she rubs their belly to get them to lie down and then trims their feet with secateurs. Tusk trimming is a job for the vet.

If you breed pigs, you are likely to encounter the odd difficulty. Our biggest drama involved a sow who rejected her first litter of three piglets. Everyone told us it was hard work rearing them ourselves, but we couldn't just 'let nature take its course', so we brought them into the house and started a routine of regular feeds. They gained weight more slowly than normal, but survived to be adopted by another sow, who farrowed five days later.

Loss of a piglet can be heartbreaking, and occasionally your sow may accidentally lie on her offspring and kill it. Avoid this by incorporating a creep area and a low bar around the walls of your farrowing house, giving the little piglets somewhere safe and protected to wait while mum settles down.

You are not allowed to bury any dead livestock on your land, instead you must call the animal undertaker (ask your vet or your local Trading Standards for a list of approved companies) who will come and collect.

As a 'fair weather' pig keeper, I have never encountered a single health problem with my pigs. Provided you buy healthy weaners from a trusted source, you're unlikely to have any worries either. Rare breeds tend to be hardy and by keeping them outdoors in a natural habitat, with plenty of space, you are providing a stress-free environment in which they'll thrive.

Good Luck!

Useful Books

The River Cottage Cookbook by Hugh Fearnley-Whittingstall
Useful chapter on pigs and what to do with the meat
The Meat Book by Hugh Fearnley-Whittingstall
Includes a handy diagram of all the cuts of pork.
A Guide to Traditional Pig Keeping by Carol Harris
My favourite reference book
Charcuterie and French Pork Cookery by Jane Grigson
If you fancy getting adventurous in the kitchen
Ham & Pigs by Paul Heiney - *History and some practical tips*
Adventures of a Bacon Curer - Maynard Davies - *Tales & recipes*

Useful Websites

www.defra.gov.uk – *Look under farming/livestock.*
www.samphireshop.co.uk – *Keep up-to-date with our pigs here at Sycamore Farm*
www.rbst.org.uk – *Rare Breed Survival Trust website*
www.britishpigs.org.uk – *To register pedigree pigs*
www.ketchums.co.uk – *For equipment 01737 812218*
www.kitchen-garden-hens.co.uk – *for other books in this series.*